My ❄ ARCTIC 1,2,3

My ARCTIC 1, 2, 3

Michael Arvaarluk Kusugak

Vladyana Krykorka

Annick Press Ltd.
Toronto • New York

The publisher wishes to thank Valerie Hatten
of the Ontario Science Centre for her assistance.

Annick Press Ltd.

We acknowledge the support of the Canada Council for the Arts, the Ontario Arts
Council, and the Government of Canada through the Book Publishing Industry
Development Program (BPIDP) for our publishing activities.

Cataloging in Publication Data
    Kusugak, Michael
      My Arctic 1, 2, 3

    ISBN 1-55037-505-9 (bound) ISBN 1-55037-504-0 (pbk.)

    1. Counting - Juvenile literature. 2. Inuit -
Juvenile literature. 3. Arctic regions - Social life
and customs - Juvenile literature. I. Krykorka,
Vladyana. II. Title. III. Title: My Arctic one, two, three

    QA113.K87 1996   j513.2'11   C96-930224-X

Distributed in Canada by: Published in the U.S.A. by Annick Press (U.S.) Ltd.
Firefly Books Ltd.   Distributed in the U.S.A. by:
66 Leek Crescent   Firefly Books (U.S.) Inc.
Richmond Hill, ON   P.O. Box 1338
L4B 1H1   Ellicott Station
                                 Buffalo, NY 14205

Printed and bound in Canada by Friesens, Altona, Manitoba.

www.annickpress.com

To the Friesen girls, Qaumaqsiaq, Angnaluaaq, Ippiksaut, Angnakuluk,
Tapaarjuk, Pallulaaq and Sarah.–M. K.

For my CANSCAIP friends, especially the
Group of Eight.–V.K.

I grew up in Repulse Bay. When I was a little boy we hunted seals, caribou and whales.

We eat seal meat. It makes your body very warm. We wear sealskin work boots called kamiik (kah-mée-eek), which are waterproof and warm. When we lived in igloos we burned the thick seal fat in our soapstone seal-oil lamps to cook and to heat our igloos with. And we used the bones for toys. Seals are very useful animals.

Caribou skins make the warmest clothing in the world. When the weather is —50° Celsius in winter, we need very warm clothing.

My favourite food is maktaaq (máhk-tah-ahk). Maktaaq is the skin of the whale. We eat the maktaaq and feed the meat to the dogs. Dogs have always been very important animals for us. We feed them very well.

We do not hunt animals all the time. Mostly, we watch them. We look at their tracks. We see how their coats change with the seasons. We watch what they hunt for food. We see how they hunt.

In this book I want to show you some of the animals we have watched and the other animals that they hunt. Watching animals is fun.

One polar bear walks along the huge floe edge
on the Hudson Bay. What do they hunt? Seals...

ᐁᑕᐅᐧᐁᐱ᙮᐀᙮'96

Two ringed seals are sunning themselves on the sea ice during the long spring day.

ᒪᕐᕉᒃ

Three members of a pod of killer whales in pursuit...

ᐱᖑᓯᔪᑦ

Four bowhead whales round the point of
Marble Island.

Five Arctic foxes look in *siksik* holes.
The pups play as the mothers hunt…

ᑕᑉᑕᒪᑉ

Six siksiks look at the foxes. Siksiks are Arctic ground squirrels.

ᐊᕐᕕᓂᕐᑦ

Seven fishermen repair a stone fishing weir...

ᒪᓯᑦᖕᓂᒃ ᐊᕐᕕᓂᖏᑦ

Eight Arctic char head up the river.

ᐱᖓᓱᓂᒃ ᐊᖃᓕᓂᖅᑦ

Nine snowy owls, beginning their hunt, take to the air...

ᕿᓗᖕᒥᕐᔪᐊᖃᑐᑦ

10

Ten lemmings scurry among the dwarf willows.

ᔕᑯᓪᑦ

A pack of twenty wolves catches a scent in the air...

A herd of one hundred caribou migrates in spring.

ᖁᓕᓄᑦ ᐊᕐᔪᕐᖏᑦ

1 000 000

Millions of berries ripen in the fall...

ᐊᒥᓲᔪᐊᓗᑦ ᓇᕝᓴᕐᑲᐅᑉᐊᖅᑭᑦᑐᑦ

One lone polar bear walks along the shore, thinking of seals.
It sees the berry pickers and says, "Never mind.
They do not look like very good meals."
It continues on its journey, looking for what it might find...

THE ARCTIC WORLD OF MICHAEL KUSUGAK AND HIS FAMILY

We live in a place called Rankin Inlet, in the territory called Nunavut. To the east of us is the Hudson Bay, which we call tariuq (táh-ree-ook), salt. To the west is the tundra, with thousands of lakes and rivers. There are no trees, no highways and no fences. There are no farms. And there are no zoos. But there are many animals.

In winter, the temperature goes down to —52° Celsius (—61° Fahrenheit). Out at sea is the huge floe edge, where the solid ice ends and the loose floes drift with the tides and currents, crashing into the solid ice and into each other. Huge pieces of ice roll over, crushing everything in their way. It is a dangerous place. It is the home of the polar bears. Here, the bears hunt their favourite meals: seals. The bears are white. Their noses are black.

One day I took my big boys, Qilak (Kee-ahk) and Ka'lak (Káh-lahk), down to the floe edge on my snowmobile and sled. We followed polar bear tracks all day long. We saw a place where the bear had caught a seal through a seal hole and had eaten it. The bear walked to a place where the ice was too thin for us, and we could not follow it anymore. Polar bears are very big but they can walk on very thin ice.

Ringed seals keep their holes open all winter long by scratching the ice with their long claws. The holes are covered with snow so that you cannot see them. In spring, when the snow on the ice melts, the holes are exposed and the seals climb onto the ice to sun themselves. The seal's skin, under its fur, is black. It catches the rays of the sun and warms the seal.

We have a telescope at our cabin. We watch seals lying on the ice. They lie on their stomachs and on their sides, their short flippers pointing up at the sky. They bounce around on their stomachs visiting each other, having a great old time. When they see or hear someone coming, they quickly slip through their holes and disappear under the ice. My little boy, Graham, who is called Ittuq (eét-took), or "old man" because he is named after my father, Kusugaq, loves to watch them.

Out at sea to the east of us is a shiny white island called Marble Island. A long time ago, many whalers came from southern Canada, Europe and the United States to hunt whales there. Sometimes you see bowhead whales and killer whales. One day we went to Marble Island in a big freighter canoe with an outboard motor. Vladyana, who draws all the pictures in my books, was scared. She thought our boat was too small to be in the middle of such a big ocean among ice floes. My boys told her not to be scared.

Arctic foxes are white, like polar bears. They have long, bushy tails. The young ones are grey and yellow. They hunt ground squirrels that are called siksiks (séek-seeks). They also like to follow polar bears, looking for scraps of seal meat.

When we were building our cabin, Inninajuk (Een-née-nah-yook), my 12-year-old son, and I watched a fox hunting siksiks across the lake. We saw it sneaking along the ground toward the siksik holes. When it got near, it ran to the siksiks, caught one and took it away in its mouth.

There is a stone weir up at the Meliadine River, just north of where we live. It is made with two rows of rock walls, built across the river like two dams, except that the water flows right through them. We trap fish in the weir, wade in the cold water up to our waists, and then spear the fish. We use a fish spear made of musk-ox horns called a kakivaak (kah-kee-vah-áhk). The water is so cold it makes our teeth chatter. Why do we do it? Because it is fun and the Arctic char is a delicious fish.

In winter, we fish through lake ice. We drill or chisel through three metres (about ten feet) of ice and jig for lake trout.

My wife, Sandy, is too chicken to wade in cold river water, but she is the best when we jig for trout through lake ice. She always catches fish.

The ukpigjuaq (oók-peegh-yoo-ahk), or snowy owl, catches lemmings and feeds them to its little ones.
Sometimes there are lots of lemmings; sometimes there are few. When there are lots of lemmings, it means there will soon be lots of snowy owls.

Lemmings are little, furry animals with short, skinny, furry tails. Inninajuk caught one and put it in a cage. It had three little ones. The little lemmings had no fur when they were born. We let them go when they could walk. They were smelly.

One day in early fall my uncle, Ussak (ooss-sahk), and I went up to Meliadine Lake. As we drove our four-wheelers along the esker, we saw a white wolf in front of the little shack there. We took our binoculars out to take a closer look. And then we saw another one, and another and another. Before we knew it there were eleven wolves. They crossed the Meliadine River and went into the hills. We followed them, but we did not see them again.

Caribou migrate to their calving ground in the spring.

Thousands of them walk together. In the fall they walk back with their calves. The calves have black faces and skinny legs. They run along behind their mothers. Even though they are little, they run very fast. They have to be fast to get away from the hungry wolves.

Our favourite time of the year is late summer, when the mosquitoes have gone, the sea looks thick and calm, and the berries and mussels are ripe. Paurngait, kigutangirnait, kingmingnait, aqpiit, kablait and siirnait all ripen and become juicy and delicious.

Paurngait (páh-oorn-gah-eet) are berries. What they are called in English I do not know. Kigutangirnait (kee-ghoo-tang-eér-nah-eet) are tiny blueberries. Kingmingnait (kéeng-meeng-nah-eet) are tiny cranberries. Aqpiit (áhk-peeh-eet) are like raspberries that are yellow when they are ripe. Kablait (káhb-lah-eet) are big berries that taste terrible. And siirnait (see-éer-nah-eet) are juicy, sour leaves. We pick them, lying on the ground for hours until our hands, faces, elbows and knees are blue and purple with berry juice and cold. This is the best time of year.

In the fall, polar bears wander up the coast, waiting for the sea ice to form. They bother our dogs; they come into our communities and scare us, because they are big and have big teeth and claws. And they are very strong and run very fast.

Last year there was a polar bear right in front of our house. Lots of people came with snowmobiles and chased it away. It was night, right after a school Christmas concert. Through our living room window we watched the snowmobiles scaring the polar bear away with their bright lights. It went away, but it will probably come back this fall.

We live in Rankin Inlet; it is the place with the big inuksugaq on the hill. There are no farms and no zoos. But there are many animals. Taima.

Some Words You've Seen in This Book:

esker–A long, narrow ridge of gravel that was formed by a retreating glacier.

floe edge–In the Arctic winter, you cannot tell where the shore ice ends and the sea ice begins. A huge shield of ice stretches from the land out to sea, and where it stops and the open water begins is called the floe edge.

inuksugak–(een-óok-soo-ghawk) A pile of enormous rocks set up to look like a man. Originally they were built to corral caribou, but now they are used as landmarks to help people find their way home.

jigging for trout–A way of ice-fishing. Instead of a rod a piece of wood, about 25 centimetres (10 inches) long with a line and a hook, is used, and the stick is "jigged" up and down over the hole in the ice.

pod– A permanent family unit of marine animals such as whales, dolphins, walruses and seals. Killer whales, also called orcas, average about 10 or 15 members to a pod. During migration groups can be as big as 250.

tundra –Areas of the Arctic where no trees will grow, only low bushes and plants. Below the top layer, the ground is permanently frozen.